Elijah De Voe

Principles of Medicine, and Medicine without Principle

Elijah De Voe

Principles of Medicine, and Medicine without Principle

ISBN/EAN: 9783744737548

Printed in Europe, USA, Canada, Australia, Japan

Cover: Foto ©berggeist007 / pixelio.de

More available books at **www.hansebooks.com**

PRINCIPLES

OF

MEDICINE,

AND

MEDICINE

WITHOUT

PRINCIPLE,

By E. DeVOE,

MEADVILLE, PENNSYLVANIA.

PRICE, FIFTY CENTS.

BUFFALO:
BAKER & JONES, PRINTERS, 220 & 222 WASHINGTON STREET.
1869.

PREFACE.

When I first entered upon the duties of a Physician, I was struck by the fact that Medical Practice of all schools was destitute of any reliable principles governing treat_ment of disease. That there was no certainty that the treatment of different physicians of any School of Medicine would be based upon the same fundamental ideas in any given case. Physicians of every school are liable to differ in regard to the expression of disease. They are also liable to differ as to treatment, even when they agree in diagnosis, i. e., in the description of disease. As I gained experience in the use of medicines I discovered certain land-marks or principles which have guided me in their application; and I have never known them to betray me.

I publish an outline of these principles, because I believe them to be of great value to the public. Although a knowledge of them may not qualify every individual always to dispense with a professional doctor, it may, and should be, a protection against medical imposition; and, furthermore, we deem it but fair that one who finds fault with the practice of others should show a better way.

It seems remarkable that principles so obvious and so demonstrable should never have been given to the world before, but it does not surprise me that they have never been set forth, advocated or practised upon by any medical sect. Being of universal application, their advocacy would not promote sectarian interest, which requires exclu-siveness. They are simple and comprehensible, and if generally adopted by physicians would soon become the property of the public. And just to the extent that medicine becomes popularized, professional doctors are flanked and driven to other vocations for a livelihood.

The reader need not be surprised when he learns that this little book will meet with no approval from sectarian doctors. It does nothing for them. So far as it has any influence it will abate their power and their incomes. How can they afford to speak well of it ? If these pages shall only serve to stimulate more popular discus_sion of medical matters, the writer will feel compensated for the trouble of their pre_paration.

INTRODUCTION.

The legitimate source of knowledge is experience. Knowledge is diffused by communication of ideas. There would be very little progress without discussion. It is plain then, reader, that your own interest should prompt you to investigate for yourself all debated questions in which your happiness and prosperity are, to any considerable extent, involved. Your health is a matter of inestimable importance to you. You may be well to-day, but only casual observation ought to convince you that you have no guarantee against the approach of disease. It is your duty to acquaint yourself with fundamental ideas of true medical art in order to protect yourself against medical imposition. Genuine rational principles of medical art are, in themselves, simple and easy of comprehension; but they are obscured and hidden beneath the verbiage and redundant learning of the schools. Many persons think it their duty to study hygiene, but almost all classes are reproachfully ignorant of appropriate methods of dealing with acute disorders. This is a deplorable truth, because it renders them incompetent to determine when they are well served. We have known a great many instances where the physician has been discharged after he had essentially conquered the disease of his patient, and another doctor called who visited longer and charged more for treating disease of his own making than the first doctor received for removing the original disorder.

To the amateur reader the nosological arrangements of disease by popular writers are very pretty, and no doubt enhance his ideas of the amount of learning exhausted in the profession—such nice classification of diseases. But when acute disease makes its appearance in a family remote from a physician, these refinements are worse than useless. They confound and discourage those who might otherwise safely undertake domestic treatment. What the people want are a few fundamental ideas that are reliable in regard to disease, and to the physiological effects of the more important medical agents. A very few medicines of the right kind will fulfil nearly all indications in acute cases. Although the human system is liable to a great many forms of acute disease, the different conditions calling for different treatment are always reducible to a few. They are not obstruse or difficult to comprehend. Men and women in all conditions in life may, if they choose, master the art of dealing with them. Every western pioneer knows how much may be done with an axe, a saw, a hammer and an augur. He also knows very well that the difference in enjoyment afforded its occupants between the most artistically constructed palace and the rude hut of the back-woodsman is not at all commensurate with the difference in expense ; and so, domestic treatment, guided by a few great principles, and availing itself mainly of domestic agents in their simplest form, would leave little to be done by "scientific practice" in acute diseases.

There is a prevalent notion to the effect that one cannot properly treat a patient without knowing what disease he is suffering from. This teaching emanates from a sect which exhausts so much learning upon the description of disease that it has little, if any, left for cure. It seems plausible, but experience proves that it is fallacious. A good medicine, judiciously used, in an advancing but indefinite ailment, nineteen times in twenty flanks and totally routs the disturbance before it reaches distinct development. It may be asked what authority we have for such treatment, the answer is, experience, experiment. We have nothing but experience and the results of experience to guide us in the treatment of disease in any stage. This serves as well in primary and indistinct stages as it does in any other. To be able to define, classify and name disease is no evidence of ability to treat it. The treatment of disease is not ruled by its specific form or expression, but by certain signs, symptoms and conditions that are more or less common to all acute disorder.

It is the object of the writer to show, first, that there are enduring principles in medicine. That these principles are simple and easy of comprehension, and that the most reliable healing arts and agents are always within the reach of the people. Second, that popular professional medicine is, in the main, artificial— the product of the misapplication of learning and science which are brought into requisition, not to aid in healing, but to obscure and nullify simple curative arts.

MEDICAL ART.

PRINCIPLES OF CURE.

Medical Art consists in the appropriate treatment of disease. It is naturally divided into two elements or sources. First, that which has for its object the study of disease; and second, that which has for its object a knowledge of healing agents, and their physiological action and influence on the human system. These two branches of medicine are entirely distinct from each other. The possession of one does not imply an acquaintance with the other. The most profound knowledge of disease carries with it no ability to cure. There . is nothing in the nature of disease that indicates its remedy. The most learned describers of disease often totally fail in the treatment of it. As healers, they are many times excelled by illiterate and unpretending herb doctors. On the other hand, the most intimate acquaintance with the medical properties, and physiological action of medical agents, does not necessarily determine their uses. One may know that this agent acts upon the bowels—another taken into the stomach causes emesis or vomiting—a third acts upon the kidneys—a fourth causes perspiration—a fifth stimulates, &c., &c. ; but he may, at the same time, be entirely ignorant as to when these actions should be set up. Some knowledge of both these branches of medical art is necessary to safe and successful practice.

The chief difficulty in the way of domestic practice, is the want of a knowledge of disease, or, more properly, an inability to judge correctly of ill conditions. For the signs and symptoms of these conditions are about all the guide we have for the treatment of acute disorder. Of the essential nature of disease, the physician knows but little more than his patron. It is an important object with medical schools to magnify this difficulty to the domestic practitioner. To that end they have made the most transcendental and hair-splitting refinements in the classification, nomenclature, description and treatment of disease. The natural effect of this is to give fictitious importance to the medical profession—discourage humble endeavors to treat disorder and to master the general principles of the healing art—and also to make the doctor more necessary. But this difficulty in judging of disease is more imaginary than real. If the reader has seen many cases of acute disorder treated by regular medical practitioners, in the early stages of the disease, he will call to mind the fact that, in most cases, when the doctor is asked to name the disturbance, he remains silent or evades the question. He does so because he knows but little, if any, more about the case than the questioner. He observes signs of acute disorder, but there is nothing in these which indicates what expression of disease will subsequently reveal itself. But we claim, emphatically, that if the doctor cannot tell what specific disorder will follow the first signs of disease, he should always be able to treat the patient advantageously. An acquaintance with the primary principles of medicine will scarcely allow him to make a serious mistake. If he treats the symptoms present appropriately, it does not signify what form of fever, inflammation, or specific disease, may follow, he will have done much to abate the malignancy of the succeeding disorder.

The manner in which most medical works are prepared for popular and professional use, indicates that the writers have very little knowledge of, or faith in, general principles as a basis of practice. The names and classification of disease they give us, imply that every different expression of it requires peculiar and exceptional treatment. So every specific fever and inflammation, as they present them to the public, has the treatment that is supposed to be appropriate to it circumstantially detailed. But this mode of dealing with the subject of disease, and its treatment, is blinding and misleading to the medical student, for the following reason: A great many different expressions of disease, which are treated of by these authors, distinctively and specifically, present essentially the same conditions of the system, and may be subdued by the same treatment.

A member of your family is attacked with acute disorder. The first thing that occurs to you is, what is the matter? What is the disease? If you knew, perhaps you could do him some good.

But you say to yourself, one is liable to a multitude of disorders, and each disease has its specific treatment; and you ask yourself, how you can be supposed to know where, in this long catalogue of disorders, you are to place the particular ailment that you now desire to treat? In answer to which, we say that it is not necessary you should know what disease it is that expresses itself by the symptoms and signs that attract your attention. It is not necessary that you should know the specific character of the disease; it is only necessary that you should be capable of judging of the conditions of the system. Almost all forms of acute disease belong to the following classes: First, the febrile condition; second, the inflammatory; third, spasmodic or nervous. Observation of disease, and the effects of medicine, have established the important fact, that all these conditions, and certain other exceptional disorders which we have not here classed, are controlled and conquered by the same medical agents and influences, whatever may be the specific name of the fever, the local expression of the inflammation, or the particular form of disturbance.

In a large proportion of acute attacks, there is some form of fever or inflammation. It has long been claimed by medical men, that fever is a general inflammation; and that an inflammation is a local fever. If these propositions are not strictly correct, they are sufficiently so for practical purposes. Appropriate treatment for one is good for all, though it may not meet in full the requirements of each particular case. Inflammation is subject to the same laws, and should be treated upon the same principles, whatever may have originated it, and whether it be located chiefly in the eyes, head, thorax, abdomen or any other locality. Having discovered the presence of fever, inflammation or acute other disturbance, these conditions are to be treated at once without regard to any special form of disease the future may develope.

All fevers, and many inflammations, are ushered in by nearly the same heralds. The premonitory symptoms of different fevers and inflammations do not differ more from one another than symptoms of the same disorder differ in different cases. Primary symptoms, as a rule, do not especially characterize the form of disorder which may be afterwards developed. Many times premonitory symptoms, generally supposed to be characteristic of typhoid fever, will be followed by transient and tractable disorders. Sometimes typhoid fever developes itself abruptly. In other cases it succeeds a long train of threatening symptoms. The warnings of diptheria and inflammation of the lungs, disorders essentially different in their action, many times correspond to symptoms often followed by continued fever. It is desirable then that the reader should be impressed with two important facts. The first is, that premonitory symptoms are not always defining—that is, they do not invariably indicate what form of disease will follow. The second is, that it is not necessary they should indicate any specific form of disease in order that the case may be treated safely, appropriately and effectively.

Science has never been able to dictate the treatment of disease which is arbitrary, or the effect of experiment. Strictly speaking, theory has but very little to do with practice in medicine. A theory may be good, and the practice which it is supposed to indicate may be bad; and the theory may be very irrational and absurd, while the practice connected with it may be successful. Although the treatment of disease is arbitrary, or the result of experiment, nevertheless a long series of experiments pointing one way, enable us to establish a rule or principle which may serve as a guide in cases which have not hitherto come under our experience. Doubtless there is some theory, more or less rational, behind all modes of treatment, which influences, to some extent, the practice of the physician,

In our opinion, disease is the effect of exhaustion, local or general. When general and slow, we may have the formation of tubercle. When general and rapid, we may have some form of nervous fever. When rapid and local, we may have a pleurisy or partial paralysis.

Pain or paralysis are among the first evidences of active disease. Intense pain is nearly equivalent to paralysis. It compels cessation of labor, or rest. The cause which produces exhaustion being removed by rest there is at once an effort on the part of the system toward recovery. Pain is sedative, and sedation abates pain. Pain relaxes—relaxation is antagonistic to pain ; thus pain tends legitimately to its own abatement. To aid its work hasten sedation and relaxation. Pain comes upon a previously healthy tissue because of nervous exhaustion ; which may follow poison, violence, great physical effort, protracted toil, exposure to cold or from wasting pleasures and excesses. Pain is the voice of nature crying out against outrage, and demanding respite. A man makes excessive demands upon his energies. He is attacked with pleurisy. Pain compels him to desist from labor, rest disposes to relaxation, and relaxation favors, as aforesaid, the subsidence of pain and inflammation. Appropriate medication and treatment expedites these processes, and when the powers of the system would otherwise fail, suitable medical treatment naturally saves life. The first object of the healer then is to place the patient in circumstances favorable to the recuperation of nervous force. In every possible way he should be prevented from the further waste of energy. If any medical treatment be of a nature to exhaust still more the vital powers, it is plain that it works in concert with the disease and helps hurry the patient to the grave. Stimulants and tonics in fevers and inflammations have the same effect as continued toil in the same condition. Rest enables the vital powers to recuperate their forces, and proper medicines give the condition of rest.

We have said in substance, that all fevers and inflammations require the same treatment. This would seem to favor the theory that disease is a unit, and appropriate treatment should always lessen the aggregate amount of disorder. It is very well known that successful physicians use fewer medicines for the treatment of acute disease the longer they practice. "I have employed

Dr. —— in my family for thirty years," said an acquaintance, "and I have observed that he almost always gives the same medicine, no matter what the disease that he is called upon to treat." This is the legitimate result of a conscientious use of proper medical agents in acute diseases. The style of the aforesaid doctor would be very frequently imitated but for the necessity of blinding the patron by the exhibition of a large number of agents. Too much simplicity would excite suspicion, whether the physician understands his business. It is, however, very easy to prepare a medicine that shall have a favorable and effective action on almost all acute attacks. A preparation that, if given in proper doses, may be used indiscriminately for every description of acute disease ; and if the quantity be adjusted to the demands of the case, no one will be injured by its effects, while in ninety-nine cases in a hundred it will be followed by beneficial results. Materia Medica affords a multitude of agents that may be used thus generally for the abatement of acute disorder. They may properly be styled universal medicines.

What are the principle characteristics of those agents for which we claim such a wide range of application ? They are emetic, Emeto Cathartic, Sedative, Diaphoretic, Diuretic and Alterative. Many single agents possess all these qualities in themselves.

In the action of this class of agents there is manifest a grand fundamental principle underlying almost all treatment for acute disorder. This principle is RELAXATION. The application of this is almost universal. · If any marked change is to be effected in the condition of the system, it must be relaxed. This fact is noticeable both in the encroachment and subsidence of disease. The system will not readily take on disease if it be in a tonic, positive condition. In a relaxed state it is susceptable to either salutary or malignant influences. If disease already exists, the administration of tonics or astringents does not assist a transition from disease to health, but tends to fix disorder. In some instances it may seem to invigorate, but the invigoration is only partial and does not remove disease. In disease the vital forces are out of balance. Relaxation of the system is necessary to the re-establishment of an equilibrium. In inflammation of the lungs a disturbing cause of the circulation of the blood is apparent. It meets with obstruction in its passage through those organs. It seems reasonable that tonics, stimulants and astringents should intensify this condition. Observation of the effects of such treatment will convince any one that it does do so. It is undoubtedly true that some patients do recover under this treatment, but this only goes to show that the inherent power of the system in some individuals is so strong as to be able to overcome great obstacles to recovery. I have had extensive experience in the treatment of inflammation of the lungs, and have always found that appropriate relaxing agents give almost instant relief. I have found this disease more tractable than any other serious inflammation or fever. The reason of this is found in the fact that the lungs are composed of a thin spongy tissue through which the whole circulation of the blood passes. Consequently proper medicines that are absorbed and taken into the circulation, are remedial by being brought at once in contact with the seat of the disease. In the hands of a large class of physicians, inflammation of the lungs is a very dangerous and intractable disorder. Their treatment violates the principle of cure, and is more dangerous because the lungs are vital organs and are easily affected, both by medicines that remove disease and those that irritate and fix it. I have, myself, many times witnessed the fatal effects of iron and other tonics and astringents given by some practitioners in this disease.

Without attempting to explain the disturbing elements in a fever the fact ·is established, by thousands of tests, that the proper administration of suitable laxative agents, under favorable circumstances is followed by a mitigation of the symptoms of the disorder. In the most obstinate forms of fever when the use of this class of agents is persisted in, the skin becomes moist and less hot, and the circulation slower and more normal. The countenance snows less anxiety, the patient is less restless, pain is abated, and if the patient had been delirious he becomes calm and rational.

One of the most prompt and efficient modes of producing general relaxation is by the use of proper nauseating and emetic medicines. Nature herself, in her processes of renovation institutes this condition, which she frequently carries to complete emesis or vomiting. A very constant symptom of the approach of acute disorder is the loss of appetite often followed by nausea and vomiting. The emetic condition of the system is antagonistic to every form of fever, inflammation and nervous derangement, and to every description of pain. The nauseant condition is purifying. The spontaneous vomitings that attend the early stages of pregnancy are natures mode of purifying the system, and serve to prepare it for the great task it has before it. I have many times demonstrated that "morning sickness" may be avoided by proper treatment ; which proves it to be a sanitary measure instituted by nature, Protracted nausea promotes a healthy action of the whole glandular system. It will unlock the most torpid and inactive condition of the liver.

In another place, under the caption Heat, we have shown the great relaxing power of that agent; the facility with which it may be applied; its wide range of usefulness; its remarkable efficacy for the removal of pain; and for the reduction of almost every form of circumscribed congestion and inflammation. Through a practice of many years, I have made extensive use of this agent, and never, in a single instance, have I known any unfavorable result follow its application.

It is my belief, based upon an extensive use of medicines, and a careful observation of their effects, that all of them which are not tonic, stimulant, or astringent, have a relaxing influence upon some organ or organs of the human body. Experience shows that the best alterative properties, especially in acute disorder, accompany medicines which are relaxing in their action, and which are most free from tonic, stimulating and astringent properties. We see the highly alterative power of simple relaxation in the application of heat. In a mental and moral sense, the benefit of relaxation is obvious to all classes of minds. It is said of men whose faculties have been long on the stretch in the pursuit of wealth or knowledge, or in the simple discharge of duty, that they need relaxation.

The simple principles of cure in acute disorder, set forth in these pages, are not given to the public as opinions, but as enduring truths that will admit of demonstration. The theory is simple, and the practice flowing from it, in the main, obvious and distinctly marked. The exceptions are few and not of a nature that are likely to confound the practitioner in regard to his duty. An oversight of these exceptions is simply an omission of the best method, and not a commission of a serious error. No harm is done to the patient. A departure from the line of conduct indicated by these fundamental ideas, I have invariably observed to be followed by bad results. Nevertheless I very well know that the practice of a large class of physicians is in direct opposition to these principles. It is well known that many persist in the use of alcoholic and narcotic stimulants, and other tonic and astringent treatment, during the most intense febrile and inflammatory action. Nothing in medicine is more demonstrable than that this treatment intensifies and protracts these conditions. Opium has a stimulating and stringent action upon the system, whether given in health or disease. The effects of its action are among the essential elements of fever. It increases arterial excitement. Even in health its administration puts the mental faculties in a delirious condition. These effects are much more pernicious when the system is already disordered. Can we hope to remove disease by adding fever to fever, delirium to delirium? The power of opium to dull sensibility to pain, renders its primary effects agreeable. There are cases where its use is pardonable and commendable—but very few where it cannot be dispensed with. The preparations of opium disturb the stomach, constipate the bowels and clog and obstruct generally, and in their reaction are very trying to the patients' nerves.

Iron, quinine, alcohol and other stimulants and tonics are made use of by a certain class of practitioners with similar unfortunate results. Their philosophy, in justification of this course, is that, in continued fevers characterized by great nervous depression, the patient must be sustained by stimulants. But the truth is, stimulants cause further nervous exhaustion, and thus increase and intensify the disorder which it is our endeavor to remove. What abates fever at once tends to the increase of nervous force. It does so because it tends to the re-establishment of function. The imperious demand of nature in fevers, especially those of a nervous character, is rest, relaxation, quietude and freedom from every description of excitement.

Genuine febrifuge agents tend directly to the abatement of the signs and symptons of inflammation. They diminish from the first arterial excitement. They calm perturbation. They dissipate that anxious expression of countenance which is one of the most universal signs of the presence of fever. They lessen pain by removing its cause, and so equalizing nervous force as entirely to control delirium.

The percentage of deaths from continued malignant fevers, under appropriate treatment, should be small. (In my practice they have not exceeded two per cent.) These fevers attack chiefly the young and middle aged, when the system will bear great strain. The writer has known a patient with delirious continued fever to recover after forty-two days of uninterrupted fever, and almost entirely without the aid of medicine—so that with good nursing, and without medical treatment, the percentage of deaths would be small, much smaller than it is under the mercurial, narcotic, stimulating treatment of "regular" practice.

It is criminal, in the light of this day, for any practicing physician not to know that there are a great variety of medicines and simples, the direct action of which is to abate fever, quiet arterial excitement, increase nervous force, and thus, as it were, to compel the subsidence of delirium and the restoration of function. These words are not uttered incoherently and recklessly, but are the deliberate expression of one who has thousands of times seen the absolute power of medicine over fevers, and knows that they may be abridged by proper treatment. I have shown, or have endeavored to show, the applicability of principles to fevers and acute inflammations. These two conditions comprehend the most numerous and dangerous expressions of acute disorder. There are few, if any, acute diseases which we know are neither of inflammatory or febrile origin. There are some which do not present to our observation any febrile or inflammatory symptoms. But it cannot be shown that these are not the result of inflammatory action. We do not know that neuralgia is not the result of inflammation in a nerve or nerves. We may reasonably presume that colic is of inflammatory origin, since it rapidly developes that condition to our observation. We do not know through what agencies convulsions, fits, tetanus and other nervous and spasmodic diseases are produced. We do know, however, that treatment which we have found so valuable in fevers and inflammations, is also appropriate and effective in all these disorders. Neuralgia, in acute cases, cannot be better treated than by such a course as will relax the whole system, and especially, if possible, the seat of the disease. The several kinds of colic which medical books describe, and all acute pain in the visceral region, are relieved by treatment instituted under these principles. And so reliable are they, that I seldom trouble myself to ascertain the exact name of the disorder, or its precise locality.

There are some diseases of the bowels which, it seems probable, are not associated with fever or inflammation. These are attended with vomiting and purging, such as cholera, cholera morbus and cholera infantum. These, however, do not constitute an exception to the rule. Experience has shown that these diseases are most successfully treated by warming and relaxing agents.

Although these principles have a general application, and are easily understood, still, a comprehension of the theory alone will not make one a skillful physician. A successful use of them requires experience in dealing with disease, study and close observation. The principle of relaxation admits of many variations ; and these must be frequently brought into requisition at the bedside. How far relaxation should be carried; how rapidly the process should be proceeded with; what particular agents are best adapted to the case in hand, are questions of art that can only be solved by trained judgment. In some cases the heat of the body should be brought down by relaxants possessing refrigerant properties. In others it should be raised by those that impart warmth. In certain cases laxative cathartics are of great utility; in others their use is inadmissable. A single case may present symptoms that require all these and other variations of treatment. And there may be twenty different expressions of disease, (different diseases,) which will all be subject to one simple application of this principle.

The reader will perceive by what has been before stated that a knowledge of these principles obviates to a great extent the necessity of determining the precise name of the disease. That nineteen times in twenty medicine may be given safely and profitably without knowing precisely what form of disorder is expressed by the particular signs and symptoms present ; and if this treatment be resorted to in the twentieth instance no harm will be done. ''Scientific medicine'' endeavors to make the public believe that each expression of disease requires specific treatment, and it thrives by the popular acceptance of this doctrine ; because it discourages domestic treatment. It is owing to the prevalence of these ideas that one doctor is spoken of as being good in scarlet fever. Another is thought to understand dealing with acute rheumatism. A third is supposed to be exceptionally qualified for treating dysentery or inflammation of lungs, but a physician who is really reliable and trustworthy in one of these disorders ought to be so in all, for all present nearly the same conditions for treatment. If a man is successful in typhoid fever and not in other forms of fever and inflammations, it is evident that his success is hap-hazard. A man who bases his treatment of disease upon the doctrine of specifics is untrustworthy. Nothing is more common than for doctors to differ in diagnosis very widely. (We have narrated a case where college professors pronounced a case of disease of the liver to be consumption of the lungs.) They may give the wrong specific, and the result might be serious. Several such acknowledged mistakes have occurred within a year past under my observation. No such un-lucky accidents will ever happen to a man who practices medicine upon sound principles.

Popular medicine is considered, even by its own patrons, as a kind of lottery that is quite as likely to give them a blank as a prize. They expect that specifics will be hurled at their diseases, which may chance to cure, or may kill the patient. "Kill or Cure" has passed into a proverb, and had its origin, no doubt, from the theory as well as the practice of "scientific" doctors. Many people believe that medicines that have power to cure are equally liable to injure or kill. Nothing can be more erroneous. But it is true that drugs that are capable of tainting the system have no healing virtues. When the vital power is insufficient for the wants of the system under the most favorable circumstances, then medicine, if it be of the right kind, will do no harm though it can be of no avail. If a man cannot walk after a friend has tried to help him to his feet, there is no reason why he should be injured by the proffered assistance.

MEDICINES.

The vegetable kingdom stands between man and the cruder elements of nature. It is the grand laboratory where inorganic substances are manipulated, softened and refined by the delicate processes of organic life, and made suitable for man's sustenance. It is this department of nature, also, which furnishes the most appropriate and reliable agents for the cure of diseases, and for the preservation of his health. The infinitely varied mingling of the nutritious and the medicinal, in the vegetable world, is a most suggestive fact. The acrid escharotic action of organic medicines in their crude form, is often softened by mucilaginous and farinacious elements. Thus, there are many organic alteratives that nourish the patient while they combat disease. So perfect and beautiful is this blending of the medicinal and the nutritive, that one may take an alterative with his dinner, and the stomach will not revolt at the compound. Indeed, these elements are so spliced into one another, that we may have any desirable percentage of either. A dish of parsnips or of water cresses, in the spring, is a genial alterative.

The power of "scientific" medicine, which has been derived from long years of governmental favor and immunity, has not sufficed wholly to destroy popular faith in simple vegetable remedies. It seems to be instinctive. The unhappy victims of scientific practice, who have been poisoned and crippled by mercurial drugs, turn with hope to the herb doctor for relief and cure. This natural bias is well understood by those itinerant medical hucksters who advertise themselves as Indian doctors. Certainly, it is a poor compliment to the light of our times, that we should look for medical treatment to the unlettered savage. That there should be a rational excuse for doing so, is an opprobrium to civilization. Plants and roots are friendly to the animal world. Brutes are liable to few diseases that instinct does not guide them to a remedy for, if they have access to woods and fields. Botanic medicines are worth as much more to man as reason is superior to instinct. By classifying and compounding them, he can reach almost every conceivable difficulty. But so conservative is nature —so loving to her children—that she favors very unintelligent uses of her products. How many millions have been greatly benefitted—how few injured—by the use of herb drinks and watery extracts of plants and roots; and of syrups made in blissful ignorance of the physiological action of the agents used; whether they were tonics or emetics, cathartics or astringents, sedatives or stimulants, refrigerants or spices, being matter of no study and of no concern.

There is scarcely an important organ in the human body liable to disease that may not be acted on by some organic agent or compound. The heart, the liver, the pancreas, the spleen, the uterus, &c.; &c., may be touched with medicine, as the pianist touches the keys of his instrument, and the response is as sure and definite.

Health consists in a harmonious exhibition of function. To express the same in other words, a person is in health when all organs in the body are performing their several duties with their normal force and vigor. In disease, there is abatement or cessation of function. Some of the organs are not doing their duty, or are doing it in part only. We hold that to be a legitimate medicine which restores and invigorates, or tends to restore and invigorate, function. Or, in other words, that is a proper medicine which restores or tends to restore harmony and equilibrium of action among all the organs of the body. The vegetable world furnishes a redundancy of plants possessing these medicinal virtues. A knowledge of only a few is requisite for the successful treatment of acute disorder. Almost every region produces abundance for the wants of its own inhabitants.

Organic medicines when appropriately used increase the amount of electricity and thus aid vital energy in its constant endeavors to expel disorder. They have peculiar power over the liver, kidneys, pancreas and the whole glandular system ; to arouse these organs from torpor and bring them to a healthy exhibition of their functions. They will not strain them by excessive stimulation or relaxation, as calomel and antimony do, nor will they remain in the system as those poisons do, causing aches and pains and thus perpetually prompting in the mind of the sufferer a curse for doctors. They will not destroy the teeth and give a putrid odor to the breath ; but they remove nausea, moisten the mouth and throat when they are persistently dry and husky. They correct the senses of taste and smell when disordered, sweeten an offensive breath, invigorate the blood, quicken the mental perceptions and perfect the general health.

It may be said with truth of organic medicines that they are natures remedies, and when properly used they attack and lessen disease, while chemical and mineral agents so freely and extensively employed being unfriendly in their essential nature to the human organism, attack the constitution itself, and by impairing its powers, augment the amount of disorder in the body.

A knowledge of the effects of medicine obviates, to a great extent, the necessity of a particular diagnosis. To one long accustomed to the use of healing agents, the manner of their reception by the system in disease, becomes an index as to the character and probable termination of the malady. When proper medical agents, fairly tried, fail to produce their characteristic effects, the hopeless nature of the disease may be inferred. Thus it is medicine becomes an instructor and communicates knowledge that could not otherwise be obtained. These facts beautifully illustrate the value of medicines and the limitation of their functions, namely :—that they are aids to vital force and facilitate its expression, but never supply its place. But inappropriate and non-medical agents blind us every way in regard to disease. Calomel, antimony, iron, arsenic, iodine and their compounds have no adjusting control over function.

Medicines which seem to have direct effect in the removal of disorders have been called specifics. This term is inappropriate, and does not fully describe their virtues They have a general action corresponding to their local effects. A medicine that has power over inflammation of the lungs, dces also remove inflammation generally. Those medicines which have the effect to reduce inflammation generally, do also overcome inflammations of a strictly local character. Only a limited number of organs in the human body can be acted upon directly through the internal administration of medicine There are no medicines which have peculiar power over inflammation in the eye, but an inflammation in this organ may be reduced with as much certainty as an inflammation in the urethra, and by the same class of agents. We have heard of specifics for scarlet fever and diptheria, but this is a gross misuse of the term. Scarlet fever and diptheria, like measles, small-pox and erysipelas typhoid fever, &c., &c., are constitutional diseases, they are all fevers and are all controlled and reduced by similar treatment.

Organic medicines can be used in combination and not have the action of one interfere with that of another. Each element in a compound produces its legitimate effect. Even those which seem to be opposite in their nature do not neutralize each other, but each exerts its own influence. Medicines of great value have within themselves apparently opposite properties, and their compound action, in some instances is highly beneficial. Rheubarb is cathartic and tonic. If that agent is not at hand its equivalent can be formed of two other agents, one of which is tonic and the other cathartic. The apparently contradictory properties of certain medicines make them especially valuable for particular uses. Here is a medicine which is emetic and tonic, two opposite properties. It is also diaphoretic and diuretic. It possesses all these qualities in a high degree. It is of great value in certain expressions of disease. But it cannot be given alone with safety, even in medicinal doses, to persons affected with disease of the heart. But, by combining it with an arterial sedative, we can qualify to any desired extent its excessive tonic action; and, at the same time, avail ourselves of its other valuable properties. In fevers we want sedation. Some sedatives reduce heat; others warm the patient. It will be perceived, therefore, that by combining and alternating medicines, we can preserve any desirable temperature.

Medicine is both preventive and curative. The developement of most diseases may be arrested at indefinite stages of their progress. This is prevention as well as cure. In arresting disease, we prevent the consequences of its unopposed developement. For this reason, it is justifiable to anticipate active disease and premedicate for its prevention. It is true many people have a dread of this use of medicine, and think it should be employed only after acute disease has been established. This suspicion of medicine, and aversion to premedication, is owing to the terrible effects of poisons so extensively substituted for medicines. It is a popular protest against what is significantly called apothecary medicine. But, in truth, no medicine pays so well as that which forestalls disease and enables the patient to evade an acute attack. In many cases malignant fevers may be averted, or favorably modified, before these disorders culminate. This is emphatically true of all diseases that have their origin in an impure state of the blood, and their name is legion. Of some of these we may predict with certainty, that they will result in the death of the patient, unless arrested by proper medication. In diseases of a scrofulous or cancerous nature, or those having a putrid tendency, the system must reach a certain degenerated condition before they express themselves locally. When taken in season, these may always be kept in abeyance by anticipatory treatment.

As a remarkable example illustrating the preventive power of medicine, we report the case of Miss Jane Matteson, of Guy's Mills, Crawford Co., Pa. Miss M. was treated in the spring and summer of 1866. Miss M. had four or five hard, rapidly growing tumors in each breast, accompanied with severe lancinating pain. When she came to me for treatment, these tumors gave every indication of being schirrous cancerous growths. Her health and strength were enfeebled. Under these circumstances we commenced medication. Within two weeks thereafter, the growth of these tumors was arrested and the pain entirely conquered. Her health and vigor have returned. Over three years have elapsed since she commenced taking medicine. Her health is good, and the tumors remain without growth or pain.

It has been our experience, many times, to be called to treat some one member of a household of six or eight individuals, for diptheria or scarlatina maligna. Apprehending that the disease would attack other members of the family, it has been our custom to anticipate it by such treatment as would, in our judgment, prepare the person for the attack, or wholly ward it off. In such cases, when the disease has extended to other members of the family, it has, without exception, been of a milder and more tractable character. We have not lost a patient of this class, though we have had the care of many such. There is always a predisposition before intense local expression. Nothing is more appropriate than premedication in anticipation of malignant disorder.

The healing power of medicine always manifests itself, when appropriately used, in the suppression of active disease, if the vital force is sufficient to support its action. Disease never stands still. It advances with more or less rapidity toward its climax, unless it receives some check which assists the system to throw it off. Some expressions of disease have a tendency to recovery—others to death. But the course of either may be modified or entirely changed by proper medicine. The eruptive fevers, such as small-pox, measles, &c., all have a tendency to recovery after they have run their course, and have reached a climax. But the system may be so influenced by medicine that their usual course may be very much shortened and favorably altered. If scarlet fever receives appropriate treatment in an early stage, it may never reach complete maturity. In many instances thorough and appropriate medication will utterly route the disorder in twenty-four hours. I have treated small-pox when the disease was essentially conquered before the eruption appeared.

A swelling, threatening abscess may be aborted, and its culmination prevented by constitutional treatment, by medication which changes the inflammatory character of the blood. I have never had a case of broken breast follow my treatment of a lying-in patient. I have been frequently warned that certain women whom I was expected to attend during confinement would have broken breasts, because on such occasions they had suffered that way before. One of the most remarkable cases of this nature was that of Mrs. —— of Meadville, the mother of four children. At the birth of each child she had been confined to the bed for several months from the effects of abscesses in the right breast. One of these had been lanced so often that the connection of the milk ducts was completely broken, so that no milk could be drawn from the breast. This woman I attended during her last confinement. Both breasts filled with milk ; one could be drawn by the child. From the other no milk could be extracted. As might be expected, the breast gave signs of inflammation, attended by constitutional fever. In this case medicine perfectly controlled both fever and inflammation until milk ceased to be secreted in the unsound breast. From this period medication was stopped, and the woman had no further trouble. It would be difficult to imagine any case that would more fully demonstrate that abscess may be prevented by medicine.

There are certain depraved conditions of the system which express themselves by general debility, by absence of energy, by swellings and abscesses, about the neck especially, chronic inflammation of the eyes or in fixed inflammation of the joints, commonly called white swellings,—hip disease, spinal affections, and that disposition of the system which accompanies the formation of tubercle. These various expressions of disease are included in the general term, scrofula, or king's evil. This affection affords an admirable illustration of that practice which relies upon the use of organic remedies. There is scarcely any expression of this many faced disorder that is not amenable to a judicious use of remedies prepared from roots and herbs. I have seen rickets in its most active stage effectually cured by a single prescription composed of organic extracts. Active hydrocephalus, or dropsy of the brain has been cured by a short course of medication. Our own practice would furnish a long and varied list of examples of cures wrought upon some form of this affection. I have also seen the corroborative testimony of others who were experienced in the use of botanic medicines. The history of medicine will not furnish a single example of any form of this affection which has been cured by mineral agents. But their use makes diseases which are counterparts of many of its expressions.

HEAT.

Is one of the most life giving and life sustaining agents in nature. As a healing principle it is of inestimable importance in the treatment of both acute and chronic disease.

The application of heat to local inflammation, to engorgements, congestions, to any part of the body afflicted with cramp or spasm, to persons suffering from sudden affections of the bowels, to any part affected with pain it can scarcely fail to give some relief, and in any acute or sudden attack no one need fear that any injury will follow the application of this agent.

A large towel or bed sheet dipped in boiling water, wrung, and applied to to the head and face for neuralgia, or to any part affected with pain, as hot as the sufferer can possibly bear it, and frequently repeated, will, nineteen times in twenty, speedily abate inflammation and lessen pain. These applications should be on a liberal scale. The cloth used should be large in order to hold a great amount of heat, and the whole head or part affected should be thoroughly enveloped. The wet cloth may be covered by dry flannels, in order to prevent the escape of heat, they may be repeated as often as the nurse feels disposed to do it.

Toothache may frequently be relieved by putting the feet in hot water and using hot applications about the head and face. Pleuritic pains, or pains in the chest, and colic pains or pains about the stomach and bowels, even when obstinate, are relieved by persistence in the use of heat applied to the parts affected. Pine and other soft wood blocks an inch thick and four or five inches square, boiled for an hour and a half in water and placed over the seat of pain, inflammation or congestion will generally be found efficacious. At all events they are good aids to other treatment. Bottles filled with hot water, heated bricks or bags of sand may be used for the same purpose. Bear in mind the object is the reduction of pain and inflammation by the use of heat. Your own invention may perhaps contrive better modes for the use of this agent.

Urine retained in the bladder by spasmodic stricture of the urethra, by enlargement of the prostrate gland, or by other causes to you unknown, will usually be set free by the use of cloths wrung from hot water and placed over the bladder, and they may also envelope the whole of the genital organs. Roasted onions applied hot may be still more efficacious, because more relaxing.

Heat to the feet will relieve pressure upon the brain. Heat to the surface of the body aids the action of the capillaries and thus equalizes and facilitates the circulation of the blood. In all these applications care should be had to keep the sufferer well covered. In all cases after the removal of hot cloths, the clothes of the sick person should not be left wet in contact with the body. Hot drinks, such as alspice, ginger, peppermint and other warming teas are effective to allay pain in the stomach and bowels. They are equally useful also after chills and other effects of exposure to cold. Injections of warm water into the bowels a little above blood heat are of great utility in obstinate attacks of colic. These injections may be repeated indefinitely or until they relieve. No evil effects need be apprehended from this treatment, even if the patient has inflammation of the bowels or of the peritoneum. Injections into the rectum for dysentery is also safe and appropriate treatment. There is no form of inflammation in which the application of heat is not beneficial.

To act reflectively and perseveringly on these few hints on the application of heat, will many times save the necessity of calling a doctor, even when not another medical agent is at command.

Heat is a relaxing agent, and is one of the many that are of universal application in acute diseases. In no instance have I known any injury to follow its use.

DISEASES OF WOMEN.

A great many books have been printed, within a few years, treating exclusively of diseases peculiar to women. Perhaps these books give additional importance to the medical profession, but it is more than doubtful if they have, in any wise, strengthened medical art. Beyond question, the practice they have originated has been detrimental to the health of the female part of community. "Scientific" medicine has endeavored to make a special study of these diseases. But its labors have served to give the subject a fictitious importance, and to open new sources of wealth and power to the profession. The medication connected with the practice is composed chiefly of fanciful chemical preparations, destitute of medical virtues, that diminish the nervous force of the patient, and thus disarm nature of her recuperative energy. New instruments are invented —new appliances are put in requisition—that result in no benefit to the patient. Where there is the remotest chance of a radical cure for falling of the womb, the use of pessaries, and other mechanical supporters, only serve the purpose of augmenting the business of professional doctors at the cost of the health and money of their patients.

Local treatment for uterine difficulties is carried to a shameful extent by a class of practitioners. Local treatment for ulcers on the neck of the womb is not only a useless but a vicious practice. They cannot be cured by local treatment; and, if they could, it would be an injury to the patient to do so. Ulcers on the neck of the womb result from inflammation of the womb itself, and their formation is nature's process for the reduction of the inflammation. To heal these ulcers by the application of caustics, or other external treatment, does not remove the inflammation which caused them; but it does lock up and fix it. If, however, we reduce the inflammation in the body of the womb, the ulcers on the neck will heal without other treatment. There are cases of uterine difficulties where it is justifiable to make a physical examination to determine the character of the disease. These cases afford a pretext for a great many abuses in practice which merit a careful consideration by thinking people. The effects of the labors of "scientific" doctors, in this branch of the medical art, is pitiful. The invalid is amused but not cured; and passes from doctor to doctor without help, until what was once only trifling weakness has become fixed disease. Thus the medical profession thrives, and thus disease prevails in the land.

Owing to sexual structure, disease expresses itself differently in the two sexes. But there is nothing exceptional in the nature of disease. It is essentially the same in man as in woman; and treatment should be conducted on the same principles. What reduces an inflammation in one will be equally efficacious in the other.

SECTARIAN MEDICINE.

TRICKS OF THE TRADE.

In primitive states of society, medicine and theology were blended. The priest and the physician were one and the same person. Medicine, like theology, had its temples, its priests, and its mystic rites. For a long time among the Romans, as well as in the early history of other great nations, medicine had no distinct status, but remained in the hands of the priests, and consisted mainly in superstitious rites and ceremonies, and in barbarous and cruel modes of treatment. We have no account, however, even in that crude state of medical art, of any agents then in use so destructive to human tissues, or so hideous in their physiological effects, as certain mineral drugs and compounds extensively used in the strong light of our day. No savage tribes of this day present such deplorable effects of vicious medication as characterize the work of our scientific doctors. The tricks of modern jugglers are professedly deceptions; but those by which doctors rob us of our money, health and life, are gravely perpetrated in the sacred name of science.

The earliest medical practitioners in this country were the clergy. The medicine man of the North American Indians is also the prophet and priest of his tribe. There has always been an effort on the part of medical practitioners, to keep medicine an occult or secret art. So we find that, among various people, and at different periods, it has been mixed with astrology, necromancy and other secret arts and practices.

The practice of medical art has seldom been sustained on its own merits—that is, by the simple, direct and undisguised skill of the physician. There is usually some appeal to the love of the marvelous, or some excitation of the superstitious biases of the invalid. He is made to believe that the doctor is an embodiment of learning and science; or, that he has some abridged method of cure—some mysterious control over disease, not bound by the immutable laws of cause and effect. In some instances, the whole art of the practitioner is summed up in the use, or the attempted use of psychological power over the patient. An example of this kind is shown in those who attempt to cure by the use of charms; by the pronouncing of certain mysterious and, to the patient, incomprehensible phrases; and by the laying on of hands. The rubbing and friction doctors, by the stimulating power of nervous force, do put in requisition an agent of positive healing virtue. In certain classes of difficulties, the effect has a degree of permanency. But, in this practice, as in those examples just named, it is chiefly the novelty and mystery of the art upon which the operator relies for business. When these cease to operate, the trick is played out.

There is a small number of practitioners who style themselves Uroscopists.

They claim to be able to describe disease, and to treat it successfully, by examining the urine of the patient, without seeing him. A sample of this is carried to the doctor in a vial, who sometimes goes through the form of examining it with a microscope, and testing it in a tube with heat and chemicals. More commonly, however, he takes the vial in hand, and putting on an air of wisdom, looks at it intently, conversing, meanwhile, with the messenger. The prescription is then put up and delivered to the agent.

These physicians rely mostly upon botanical agents, and it is but justice to admit that they are far more successful in the treatment of either acute or chronic disease than the self-styled regulars; as any physician must be who uses organic remedies with only indifferent skill. Their practice is not so objectionable as their pretension to arts they do not possess. The Uroscopist sees no more in the urine than any other doctor, and it is not from knowledge obtained from this source that he is enabled to judge of the patients disease, or what medication it calls for. If he knew definitely the constituents of the urine, its elements, and the precise quantitiy of each element, it would still afford him slight knowledge of the patients disease, and no clue whatever to treatment. That they often help their patients without seeing them, there is no rational doubt. A little knowledge in regard to the condition of the patient can always be drawn from the messenger, without asking a question that shall appear to him as having any bearing on the case. If the doctor can only ascertain whether the disorder is acute or chronic, which, in ninety-nine cases in a hundred he can easily do, he can treat him safely, and in a majority of instances, successfully. This goes to prove what we have said elsewhere, that professional medicine is in large part artificial. The pretended study of the urine is a blind. The patients attention must be diverted from the simple philosophy of cure, and fixed upon some imaginary art or science of great profundity.

An examination of the pulse was once considered sufficient for the discovery and treatment of all diseases that flesh is heir to. It is still of more value in determining the condition of the patient than any other one sign we have. But it would be useless for any man in these days to pretend to base a practice upon an examination of the pulse alone. A knowledge of its value is so wide spread and so demonstrable that but few people could be humbugged by any extraordinary pretensions to science in that direction. The urine is sometimes an aid in the study of disease, but it is ridiculous to suppose that a man can learn more from a bottle of urine alone than he can by systematic methods of inquiry and physical examination.

The Clairvoyant would have his victims believe that he can see better through the top of his head than through the natural organs of vision. Mrs. A—— says a clairvoyant whom she employed described her disease "exactly in every particular." "And did you yourself know your disease, when you called him in counsel?" "No, of course not," she replied, "or I should not have sent for him." "Well, if you did not know what your disease was, and your counsel has described parts that cannot be examined by aid of any or all the organs of sense, how do you know that he told you the truth." Of course there could be no satisfactory answer to the question. It is easy to make statements in regard to organs in the interior of the body, for what is affirmed cannot be disproved. It is characteristic of many modes of practice, to tempt the confidence of the patient for the purpose of securing patronage by endeavoring to show a knowledge of his disease, but as we have elsewhere said, a knowledge of disease is only one branch of medical art, and does not imply ability to cure.

The artful medium uses departed spirits to unfold the marvels of disease and cure. Many of the mediums are young, and confessedly not experienced in medical matters; but they affect to be acted on by higher and more intelligent powers. No doubt many of the patrons of uroscopy, would consider these latter as humbugs. But the trick has the same significancy in all. It is a mode of creating prestige—a means of securing confidence, and is essentially the same as that which gives importance to all sects.

Homeopathy is the shadow of Allopathy. Its birth was aided by a knowledge of the disease-making effects of their drugs. No man ignorant of allopathy would ever have originated homeopathy. The principal trick of this sect, unlike the examples just given, does not consist in the pretense of a marvelous manner of detecting disease, but in the principle upon which they administer medicine and discover the remedies of disease. It has, however, other tricks by which it bids for patronage. Among the most prominent are, its claims to be scientific—its doctrine of specifics, with its multitude of agents, and its strictly exclusive character. This doctrine of specifics, that is, that certain medicines have exclusive or exceptional power over particular expressions of disease, is erroneous. There are no specifics in medicine, no medicines that have exclusive power over any particular expression of disease. A proper use of the same agents are effective with all forms of fever, and all local inflammations. Pleurisy root abates pain arising from inflammation of the pleural membrane; but it has no specific or exceptional action in this disease. It abates pain arising from fever or inflammation in any part of the body. Besides, there may be a hundred medicines besides pleurisy root that conquers pleurisy, and acts in the direction of cure over all inflammations and fevers. Colchicum has some power over acute rheumatism and gout, but it has the same power over other inflammations and fevers as far as its use is called for, On the other hand there are many other agents which have far more power over gout and rheumatism than colchicum. Some medicines act with greater energy on certain organs than on others, but their beneficial effects on the system are by no means bounded by their local action. They reduce inflammation located in other parts of the body, as effectively as those which express themselves mainly in those organs which seem more especially acted on by the medicine. For instance some medicines act with exceptional energy on the kidneys, but the same power these agents exert over inflammation in these organs, they would manifest over inflammations in other organs. By their action on the kidneys, they energize their functions, and cause certain excretory changes in the blood. The blood so changed, exerts a healing influence over all affected parts of the body.

Homeopathy may properly be called a fancy fraud of scientific medicine, and, next to Allopathy, is the best planned and most thoroughly ramified system of humbuggery that is any where trampling upon and smothering the simple healing arts. No doubt Hahnemann was disgusted with the systematic disease making of Allopathy. He tells us that the disorders made by its drugs are the most obstinate of all chronic affections. But, singularly enough, the chief poisons of these disease makers are Homeopathic specifics. Like many Allopaths of his day, most likely Hahnemann was a medical skeptic; and, in breaking away from the old school, may have said to himself: "I will beat a hole in the noisy drum of these disease makers. I have no more faith in medicine than they have, but my conscience will not let me make so much disease: I will bring a rational philosophy to the aid of my ghost of medication; I will tickle the imagination of the lovers of the marvelous by my extreme dilutions. My sugar-coated pills will be a pleasing contrast to the disgusting doses of the Allopaths, and I shall win." Homeopathy is winning. In many large cities it is driving Allopathy to the wall. Hence the hostility of the latter to Homeopathy exhibits the very essence of malignity.

In every city there is a percentage of people who want to be doctored. They don't expect to be cured, for the reason that they have no positive ailments except ennui. They don't want to be out of fashion; and it is fashionable to receive the visits of a physician. Besides, they must be amused; but they don't want to be poisoned—that is, but little; and they want that little concealed in sugar pills. Little pills that will not rack the nerves of those that swallow them. Homeopathy is admirably adapted to answer this demand. As a fancy humbug it is superior to Allopathy, and better adapted to the wants of our civilization. No sect can become popular that does not practice some form of imposition. Practically the Homeopathy of Hahnemann institution is defunct. Extreme dilution is repudiated whenever the case in hand calls for vigorous treatment. In fact we have seen the Homeopathist advertise his medicine as concentrated, in comparison with the weaker and watery preparations of other physicians.

It is to the interest of all medical sects, that the people shall believe that the medical art is beyond their comprehension—at least without years of preparatory study. It is true that their arts are hard to understand. They are artfully made so. Physicians complicate disease, and complicate treatment. But genuine medical art is very simple. If a man has fever, and you know how to abate that fever, you know considerable medical art. And this you may easily learn to do, though you cannot write your name or read the alphabet.

The public mind must be confounded by involved and contradictory theories, and absurd practice. To make their teaching consistent, plain and truthful, would be self-destruction. It would be no less than to tell every one how to do without their services. To fully develope and empower an orthodox sect, the public must be overwhelmed with awe by the psychological power of learning. To this end the students brain is filled with technicalities and useless jargon. Great importance is attached to qualifications and attainments entirely irrelevant to healing arts. A medical phraseology is invented for effect—utterly without utility, except to excite wonder for learning as useless as it is incomprehensible.

For illustration, let us show how the word Neuralgia has served the profession in this direction. A man is severely distressed in the region of the stomach. He says to his physician, after describing his case: "What is the matter with me?" "O," says the doctor, "you have neuralgia of the stomach." The patient is now satisfied that the doctor profoundly understands his case, and receives treatment with confidence. If the doctor, in answer to his patient's interrogation, had said, "you have pain the stomach," the patient would very likely have replied, "I know that as well as you do." But, practically, the doctor told him no more when he called it neuralgia. Neuralgia reveals no more to the doctor than the word pain does to the patient. There can be no pain without disorder in some one or more of the nerves of sensation. It is true that when the word neuralgia was first introduced into medical literature, it signified pain in a nerve independent of other organs or disease. But the definition was soon found not to be strictly correct. It is scarcely possible for a nerve of sensation to be affected independent of other organs. This disorder is corrected by treating it through the medium of other organs. Literally neuralgia is pain in a nerve. But all pain is pain in a nerve or nerves. The nerves only can make one conscious of pain. But pain is a very common sensation. It is so common as to be esteemed vulgar by the "regulars." To tell a man he has pain, is to tell him nothing new. So a scientific doctor calls it neuralgia. Neuralgia is science, and nobody is supposed to comprehend it but a doctor.

The institution of the family physician is an outgrowth of a disguised and artificial profession. It is, in fact, a relict of the dark ages, when the priest and the physician were one and the same person, and did the thinking, theologically and medically, for all classes; when to differ with the opinion of the doctor in regard to health and the proper treatment of disease, would have been deemed as presumptuous as to differ with the priest about matters pertaining to the soul. The institution is not in accordance with the spirit of our times, and is a bar to the diffusion of medical knowledge among the people. It is the interest of the family physician to discourage the introduction of knowledge which would better enable his patients to dispense with his services. When a family has found a physician whose skill, integrity, and prompt attention to business, commend him to their patronage, it is proper that such family should employ him when they need his services, *for what he is best qualified to do.* But what is there in medical art that should exempt the physician more than men in other vocations from the laws of competition? Nothing. Only an outline knowledge of the principles of medical art would make this fact apparent. Most people confine their patronage to one physician because of their acknowledged inability to judge of the value of medical service. Long acquaintance with one man is the best guarantee they have for good service. They think they know him. But it is an essential qualification of a "scientific" doctor not to be known. He wears perpetually a professional mask, as impenetrable as that of the veiled prophet of Khorassen, to the unmedical patron. This mask is an integral part of his professional qualification. It is studied and taught systematically. The Clarion County Medical Society of the State of Pennsylvania, in its code of ethics published in 1867, says: "A peculiar reserve must be maintained by physicians in regard to professional matters." That reserve is intended, no doubt, to conceal from those whom this society, in high church style, call "laymen," professional ignorance, and that laugh in the sleeve which must so often reveal itself to the members of the society. "My son," said a considerate father to a child whom Providence had somewhat stinted in mental endowments, "if you go with me to town you must not talk, and people will not know you are a fool." The sharp business tact of the American people is less conspicuous in their dealings with the doctor than in any other department of human affairs. One would suppose that life and health were of sufficient value to stimulate cautiousness and conservatism to the last degree. But it is astonishing how easily men and women, in all conditions of life, can be imposed on in medical matters, when they are approached with the artificial authority of scientific medicine.

The "regular" doctor is generally opposed to ordinary modes of advertising. He is so for these reasons:—If a regular, combined and associated physician can impress the public with the idea that the representatives of his school only are properly qualified physicians, they have no occasion to advertize in any other way, nor to strive for success as healers. They are sure to do all the chance business and as much more as they can make by the use of their drugs. They understand very well that, if they advertised their cures, they would be estimated and measured by their success in healing. "Don't talk about cures," said a "regular" college professor to his class, "but base your claim to patronage on the fact that you are a regular physician." "The people," said he, "won't reason on these matters. They can't. You must command them by authority—that always controls the unreflecting; and then it little matters what your work is, it will be considered right." No! the regular physician does not advertise directly; and considering the deadly effect of his drugs, and the terrible character of his work, it is a great misfortune to the public that he does not.

But if the "regular" does not advertise, as aforesaid, he has, nevertheless, a very effectual mode of blowing his own trumpet. Affecting to contemn the judgement of the public as incompetent to judge of the character of his work, he looks to the profession for his honors. The very assumption that the calomel schools only graduate regular bred physicians, is a mode of advertising. Their affectation of mystery, and their endlessly tautological phraseology, based upon the most inconsequential learning, is still another mode of advertising.

Another mode in which " scientific" popular medicine uses its artificial prestige and power to sustain itself at the expense of independent thought and action, and to appear wise in the public eye, is to apply some epithet of reproach to stigmatize all who shall attempt to practice medicine without its seal and authority. This epithet is quack. With your scientific doctor, every kind of medical practice is quackery except his own. If the best representatives of scientific medicine should get progressive ideas, denounce the authority of their school, and set up as independent practitioners, the immaculate fraternity of regulars would call them quacks, as they do all who defect from their ranks. But these men would know just as much outside of the calomel school as in it. So reader, you see that quackery does not signify want of qualification, but want of a particular kind of authority, in other words the quack is a medical heretic who presumes to do some of his own thinking. Now, let us suppose a young parchmented "regular" fresh from the schools, commencing practice under the auspices of the fraternity. His mouth is full of technical phraseology and the jargon of scientific medicine, yet he may lack all the instincts and philosophy of nature's physician. He may be utterly incapable of comprehending the higher principles of the profession. He may poison his patients with calomel, corrosive sublimate, antimony, &c., &c. He may mark his ride everywhere with wrecked systems, the legitimate effects of his accursed drugs. He may fill his pockets with gold, and give his patients in return shattered constitutions and depraved appetites. This man has a diploma, and in the language of the schools, is a "properly qualified physician." Those who strain their lungs in crying quack, would never apply the epithet to such a case as this. So reader, when you hear one denounced as a quack, reflect a moment—consider who uses the word—what is its signification, and you will not be disturbed by it, quack! quackery! ! Why, that quackery must to be deprecated—most abundant, is licensed quackery—authorized quackery—popular quackery—quackery that maims, disables and destroys, all in the name of science.

This is the spirit of regular organized popular medicine. To make disease, and therefore business for itself. To compel public patronage by combinations. To crush out independent practitioners by refusing to counsel with them, and to stifle all criticisms and independent thought in its patrons. "Any man" says the great Harvey, "who presumes to dispute their doctrines, or to practice in oppositions to the prescriptions based upon them, is denounced as a quack and a murderer, and visited with a malicious persecution that stops at nothing short of destruction, root and branch."

The prestige and power of the calomel schools depend chiefly upon artificial aids. If one could take from regular organized medicine that "peculiar reserve" which its representatives are instructed to maintain towards those whom they call laymen—if one could take from it its "fuss and feathers," its affected dignity, its inflated phraseology, its persistent time-honored humbuggery, and its disease-making drugs, the effect would be like taking gas from a balloon, the institution would collapse. Where five doctors now get rich, one would scarcely get an economical living.

SECTARIAN MEDICAL LITERATURE.

A celebrated French surgeon, who had the benefit of forty years' experience in the armies and hospitals of France, said that if all the medical literature then extant were to be swept away, nothing would be lost that was worth preserving. Whatever may have inspired such an opinion in his day, it is quite evident that the floods of medical trash with which we are inundated, in our times, would much better justify such judgment. These books show conclusively that it is very difficult to convey to the minds of others distinct ideas of those things which the writers do not understand themselves. It is probable that most medical books are intended by their authors as nothing more than professional advertisements. Others are written for the purpose of introducing new humbugs as the old ones begin to play out. It is notorious that none of them have communicated to the public knowledge of practical utility in regard to the treatment of disease. The most careful analytical revision of all the medical literature extant, of every sect and school, would not furnish one healing principle that would be accepted as such by all classes of physicians. No agreement between physicians, even of the same school, consulted separately, can be counted on in their treatment of the most common expression of dangerous acute disease. Why is this? Is there nothing tangible and permanent in medicine? Undoubtedly there are principles in medicine as permanent as natural organic law. But it is to the interest of all medical sects that no such uniform principles of medical practice shall be established. Most medical books are written by the representatives of some school or sect, and who are, consequently, interested in disseminating exclusive theories. There can be no exclusiveness where every practitioner is governed by the same principles. Beside, the treatment of acute disease is of such a nature that, if uniform principles of the right character were adopted by physicians, they would soon become the property of the public, and few professional doctors could be maintained. Thus, it is plain, that all sects are interested in confounding the common understanding in regard to the most fitting modes of medical treatment. Whatever common experience approves, the dictum of the schools condemns. That, therefore, is a good professional book which, while it seems to mean something, conveys no knowledge of practical utility to the reader. These books that claim to be scientific works, and this pretension to science as an indispensable requisite to medical practice, is the wool that obscures the mental vision of so many thousands of people, and makes them willing to be poisoned, crippled and maimed.

It is amusing to hear medical dogberrys discourse of the learning of physicians. They say that "Dr. A. is very scientific." "Dr. B. is one of the best read men in the country." Why, if a man were to read to any considerable extent the medical literature of the day, under the impression that it would convey solid, useful knowledge, he would be in great danger of becoming a lunatic or an idiot. Good practical physicians seldom have many medical books, and the few they do have are but little read. Practical dealing with disease soon drives away the mists, contradictions and absurdities so abundant in medical literature.

Much is said about different systems of medicine. There is no 'system in medicine. Medical Art is not the result of the demonstrations of science, but of experiment pure and simple. It would be just as fitting to speak of a system of sculpture, and a system of fiddling, as a system of medicine. If there were an established system in medicine, the doctor's work could be pre-determined. Though medical literature is abundant as the locusts that plagued the Egyptians, the duty of the physician can only be indicated by the occasion and circumstances that demand his services. Doctors disagree! In the intensified language of the day, that is no name for it. In the same case one physician vomits where another expresses the positive opinion that emetics are contra indicated. This one proposes a cathartic; that one declares that physic would needlessly weaken the patient, and it is his especial aim to prevent a movement of the bowels. A fourth insists upon the application of blisters or antimonial sores. The fifth says it is outrageous practice to interfere with the integrity of the skin, and to take away that strength which the patient needs to throw the disease. This doctor does wonders by sweats. Is confident that he could sweat out a typhoid. The next M.D. knows sweating weakens a patient. He don't know it does anything else.

Practical medical truth is buried under the verbiage of the schools. The instruction of popular medical works is an *ignus fatuus* that leads to bewilder, and dazzles to blind. The precepts of medical literature shock both common sense and philosophy. The great work of the medical teacher of to-day is to unlearn the public of its errors. Men have so long helplessly leaned upon doctors, that they seem almost incapable of independent thought in medical matters. Such is the power of authority made by schools and associations, and utterly without the support of science, that it induces thousands daily to submit to treatment outrageous to reason and destructive to vital power. This wonderful psychological influence folds in its embrace not only the ignorant and illiterate, but men of intellect who have devoted their lives to literary pursuits. 'Tis true, 'tis pity! and pity 'tis, 'tis true!

SCIENTIFIC DOCTORS NOT HEALERS.

The flourishing state of the medical profession, the great number of practitioners that rapidly make fortunes by doctoring, is demonstrative proof that popular medicine has little in common with the healing art. If the labors of these doctors were exclusively devoted to the removal of disease, the effect would be manifest in their diminished and diminishing numbers. Their work should continually abate the aggregate amount of disease and consequently lessen the number of practitioners as it reduces the source of their subsistence. If popular medical sects would give positive evidence of skill in the healing art, their work should constantly tend to make their own services less in demand. Skill in healing is self nullification. That is the best doctor in any community, who by his wisdom and art so preserves the health of his patrons as to make his services unnecessary or at least rarely in demand. If by his medicine and instruction in a single visit a doctor aborts a fever, or breaks down an inflammation that would otherwise have run thirty days, he gets only one thirtieth of the recompense of the unskillful physician whose services would be required for as many as thirty visits. It is a fair presumption then, and a just estimate of ability, that of a given number of doctors who have an equal number of patrons under equally favorable circumstances, that is the best and most skilful one who has the least to do. The best evidence then which the representatives of popular medicine could furnish of their power over disease, would be to make their services unnecessary to the public, if not absolutely yet proximately. It is self-evident that such is the tendency of medical skill.

Proper treatment of disease is not only curative but hygenic. To remove the source of disease in the most simple and effective manner is to lay the foundation of health. Usually such treatment conveys valuable and saving instruction to the patient. A good physician, from the very nature of his work, will so disseminate hygenic principles as to continually diminish the medical requirements of his patients. The highest skill in the physician is to abbreviate disease and to prevent what would otherwise have culminated in malignant disorder. But this skill is rarely appreciated or compensated. The patient does not know what he has been saved from, and cannot, therefore, properly estimate the services rendered. Doctors who have many patients who are very sick, and sick for a long time, get both reputation and money.

Healing is very unremunerative. It is disease-making and protracting that pays. We had a patient who paid us a certain sum of money for sixty days medical treatment. At the end of forty-five days he asked to have one-fourth of the money refunded, on the plea that he was completely restored to health. Had he not been entirely cured at the end of sixty days, he would not have been dissatisfied with the investment of the whole sum.

Organized or sectarian medicine does not depend solely upon natural or fortuitous causes of disease for business ; but by its own drugs, combinations and public teachings it puts causes in operation that make employment for its own practitioners. An individual is doctored with calomel and opium for the first time. This treatment is the seed of perrenial diseases. From that time the profession has a heavy mortgage on his earnings, from which it draws compound interest. He is doctored into disease with calomel and other poisons, and calomel is still used to doctor the disorder it has made.

Calomel when taken into the system in health or disease produces the signs and symptoms of a malignant inflammation, and that of a far more obstinate and persistent character than most spontaneous inflammations. Opium is a powerful stimulant astringent that increases arterial excitement and causes a kind of delirium even in health. Iron and quinine are powerful clogging tonics, and therefore every way adapted to the aggravation of febrile and inflammatory action; yet the use of these agents in many forms of fever and inflammation, is authorized by the teachings of the allopathic schools. At the same time they repudiate those simple febrifuge medicines that are always direct and effective in their action to abate fever and inflammation, and never do any harm. Thus it is that "scientific" medicine confounds the common understanding in regard to the action of medical agents, and debauches the public mind in regard to disease. It condems what heals and saves and authorizes what makes, intensifies and protracts disorder. Its self-created authority destroys self reliance among the people, and discourages domestic treatment.

We have before remarked that scientific medicine is largely artificial, ingeniously creating a demand for its own services. A community possessing only a limited acquaintance with the simple rules of healing, at once annihilates medicine as a distinct profession. For instance, the Shakers maintain no professional doctors. They claim to have had but one case of typhoid fever in thirty years, and that they say was the result of oversight, and is not likely to happen again. They say, sensibly enough, how absurd it is to go to a doctor for the recovery of your health, who is interested in keeping you sick. Other communities have demonstrated the uselessness of an organized medical profession. Repeatedly in my practice I have known a calomel doctor employed through the influence of fear and ignorance to visit a patient weeks, and in one instance sixty days after my treatment had reduced every manifestation of disease except what the doctor made for himself.

It is a curious fact that skill in healing is incompatible with distinction to its possessor. The most skillful have the least to do. Healing is a negative art. It consists chiefly in arresting and preventing disease. Unlike painting, sculpture and music it can leave no enduring monuments of the artists work. No healer in modern times has achieved cosmopolitan eminence or any considerable fortune; and it is highly probable from the nature of medical art that the reputation of Esculapius, Hippocrates and Galen was owing to other causes than to their healing powers. The reputation of all sectarian doctors is mostly local and purely artificial or professional—made for themselves by themselves, on the principal of mutual admiration and eulogy. Jenner, for his application of vaccine may be thought an exception. But his discovery was preventive rather than positively healing. And his case furnishes another demonstration of the hostility of the profession to progressive medicine. He experienced at the hands of the profession all that malice and envy could inflict. It resisted his discovery as long as resistance could avail. Harvey's discovery of the circulation of the blood, although it was purely scientific and not directly connected with the healing art, met with the same opposition.

The healing art is too simple in its principles to admit of its practitioners acquiring an extensive reputation. Dr. Dixon, of New York, has attained greater notoriety by his profound skepticism of the efficacy of medicine, than has attached to any healer of our times. All that has been established in regard to the treatment of acute disease may be comprehended and practised by men and women of ordinary capacity. That these simple arts are not better understood is a reproach to the people of the nineteenth century, though this want of knowledge is undoubtedly due in part to systematic efforts of professional medicine to keep the public ignorant. Domestic medical art is sneered at, condemned and repudiated by the scientific practitioner, whose fortune depends upon his ability to repress everything like self reliance among its patrons.

The art of reading is not so simple as that which relates to the management of acute disease. Half the amount given to doctors would pay for instructing every head of a family in the action of the most important medical agents, and also when and how to use them. Only feeble light in this direction would enable the patron of the physician to detect and expose those very superficial arts by which a fever or inflammation is sometimes protracted five or eight weeks, that could be subdued in as many days by appropriate treatment. There are cases occasionally that require close and long attention from the physician. These serve as an effectual blind to the patron who being ignorant of the principles or rules governing medical art, cannot tell when these occur.

All medical sects are opposed to the healing art—at least so far as their organized and combined action is concerned. No sect or organization works for its own destruction. The vitality of medical sects is in disease. The more disease the better they thrive ; the aggregated influence of association cannot promote cure when the tendency of cure is to the destruction of the association.

What man in a race is going to run fast when it is the hindmost who wins the stakes. In such organization there is a tendency to the exclusion of the most reliable curative agents, and to the introduction of those which make and prolong disorder. Like the manna which fed the children of Israel in the wilderness, the agents of cure are scattered broadcast by a bountiful providence over the whole earth ; and like that manna too, they perish by every attempt to hoard or monopolize them. There are many enterprizes in the affairs of life in which co-operation, association and organization are commendable because the interests of the association and those of the public are identical. It is quite the reverse in medicine, which thrives upon the misfortunes of its patrons. Medical organization, therefore, is stimulated by its own interests to increase and prolong disease, which is the source of the power and wealth of its members.

Thus one can see plainly enough that healing is a positive damage to the medical profession. The very excellency of the healers work makes him hostile to their interests.

In order to confound the judgment of unmedical persons, and to excite profound reverence for *science*, the profession sometimes authorizes the most absurd treatment, such as the application of ice to sufferers from asiatic cholera; and the use of calomel, and alcohol and narcotic stimulants for the reduction of delirious nervous fevers. "What is the remedy in this case?" inquired a student of his tutor. "The remedy," said the old doctor, "is warm water, but the practice is cold water." How could the remedy be one thing and the practice another? Very easily. The schools and associations may establish a practice, but experience determines the remedy. The schools and the associations are interested in the prosperity of doctors—they are not equally interested in the removal of disease. So extensively has the popular judgment been perverted by the mineral practice, that many people think it necessary that, in order to be made better, one must first be made worse. Whereas the curative effect of a true medicine is direct from the first. With the best treatment, fevers will occasionally find their victims in early and middle life. But, in enlightened countries, where we have the benefit of scientific medicine, a much greater number are killed by the doctor's drugs.

Mrs. P. called upon me and said she had just got out of doors after a long run of fever that, she said, had cost her upwards of seventy dollars, though she lived within a short walk of her doctor. She wanted to know if I could do something for her towards the recovery of her health. It was plain to see that she had once possessed a superb physical system, and it was sad to note what change had been wrought by seventy dollars worth of scientific doctoring. "I had," said she, "when I was taken sick, beautiful teeth—look at them now;" and her eyes filled with tears. Several of her teeth had fallen out and the remaining ones were black and loose. Her nerves were shattered and made susceptible to every change of atmosphere, and her whole system without tone or recuperative power. If there is any time, in the history of an individual, when he should be free from disease—when he should least need medicine—it is undoubtedly just after a fever properly treated. Malignant fevers do indeed temporarily impair energy; but, when they do not take life, they do not taint and disorganize like the calomel doctor's drugs.

Mr. O. lives in a neighboring town of Crawford Co. Said he was thrown from a horse and had two of his ribs broken. Having a wholesome dread of doctors he determined, if possible, to do without their services. Two weeks had passed and he had so far recovered that he could get out doors and direct matters on his farm. He was every day improving; but, in an evil hour, was persuaded to send for a doctor. The first drugs he took brought him to his bed, where his ingenious doctor contrived to keep him for four long months. From this bed he arose with the loss of most of his teeth (by calomel), and the permanent impairment of what had been a strong, well-balanced physical system. His doctor bill amounted to one hundred and fifty dollars, to say nothing of the damage he sustained by being unable to direct his affairs. This was "scientific" doctoring; and it is only by the aid of the sacred name of science that men are enabled to practice such barefaced villainy.

"The apothecary doctors," said a plain spoken man from the country, "charged me over one hundred dollars for killing my wife with calomel. I am myself poisoned by the infernal drug, and never expect to recover from the effects of it."

A large percentage of the reform doctor's time is devoted to the renovation of systems racked, tainted and prostrated by mercurial treatment. These human wrecks are to be found on every hand, though the number is scarcely suspected, except by those who are called on to undertake the difficult task of rebuilding on the ruins.

"I have come a long distance," said a woman of twenty-five years, "to see if you can do anything for me," and she showed two crippled hands and arms. The left hand was stiffened at the wrist, and the fingers so contracted as to render the hand useless. The right arm was adhered at the elbow joint, and stiff. The palm of the hand was turned down, the two bones of the arm being crossed and immovable, and the whole limb without function. "I had Typhoid fever, and my doctor, who is a regular bred physician, said the fever took a bad turn."

"This is the work of poison, Madame, given by your regular physician. Typhoid fever may kill, but it never produces such effects as these."

Thus is poor humanity tortured and crippled. Every considerate person must know that drugs which produce such terrible results locally, must shatter and deprave the whole system. And so it was in this case. This unfortunate woman was not only deprived of the use of her hands, but had other ailments, the effect of poisonous treatment of the gravest character, such as she could be redeemed from by no medical skill.

Mr. W. is blind in one eye, and nearly so in the other. This is his story:— "I was living on a homestead which I had earned by hard work. I was worth two thousand dollars. I took an inflammation in one eye and applied to a regular physician. His treatment increased the malignancy of the disease. I went to New York to a *more* "scientific" doctor, and within one month I lost entirely the sight of an eye, and the disease had attacked the other. I doctored on 'till I am as you see, nearly blind and a beggar." The result would have been far, far better had he not went near any physician. The tendency of inflammatory action is to abatement. With quietude, spare diet, and good nursing, he could scarcely have suffered permanent impairment of his eyesight.

Many cases of fever especially of that class called nervous, beyond a peradventure, are made by certain modes of doctoring. We were called to see a lady in Meadville. "Dr." said she, "what is the matter with me?" "You have fever, madam." "Yes, but what kind of fever," she inquired. "That is more difficult to tell. It has not the appearance of any legitimate fever with which I am acquainted." The patient then informed me that a short time before she had been terribly salivated by calomel and had not been well since. Further investigation showed that this fever was a secondary mercurial fever, caused by relicts of mercurial compounds still in the system ; and this disease-making practice is scientific medicine !

In the city of Buffalo I was called to see a young man. We were requested to examine the patient and report to his friends in an adjoining room. "Well doctor, what is the matter with him." "He has a bilious attack," we replied. "A bilious attack, indeed! why, a prominent surgeon and practitioner of the city has been called to attend him and says he has typhoid fever. That he will probably be sick four or five weeks at least. He has prescribed calomel to be followed by milk punch, with Dover powders at night." "I can assure you that the young man has no typhoid fever, but the treatment you have named, if continued, will be pretty sure to make one." A fever would be generated in any well person subjected to such treatment. At the urgent request of the friends we took charge of the patient, visited him four times, and three days later saw him down town going to settle with his "scientific" doctor.

Such developements would be much more frequent but that "scientific" medicine has such a grip upon the public mind that the victimized patient once entangled in the meshes rarely escapes, but is doctored on to the bitter end. One of the saws by which this influence is maintained is that "It will not do to change medicine." The doctor may lose patient after patient. Those that still live may be hourly sinking under the effects of their poisonous, disease-making drugs, and when the confidence of the friends of the sick in the skill of the doctor is giving way, they are told that it will not do to change medicine, and thus thousands have been deluded into the acceptance of this absurd dogma. That any person having a modicum of common sense should be influenced by such a doctrine shows how blindly and slavishly the maxims of medical auto-crats are accepted by their patrons. The most limited understanding should be able to comprehend the propriety of discontinuing the use of drugs which are not reaching the case, but which are making new disease. If a man's hand is in the fire, how much argument should be required to convince him of the propriety of withdrawing it. Ah! how profound are the precepts of "scien-tific" medicine.

The most popular medical sect, that which has most immunities and receives most governmental favors—that in which the practitioner most rapidly acquires a fortune—is also the most corrupt—is least successful in healing—and as might be expected, has least faith in the curative power of medicine. Sects become rich and powerful in proportion as healing arts are ignored and laid aside, and their place supplied by disease-making drugs.

Doctors, like other people, are governed by motive. The doctor who works in combination—that secures business by the authority of his association and not by his skill, sees with his eyes half open that his mineral drugs and chemi-cal poisons make him more business than medicines. He therefore easily learns to forget the latter and to rely on the former, especially as he is compelled in a measure to use and recommend them by the terms of his fellowship in the med-ical fraternity. "Come with us," said a member of one of those combinations when we first came to Meadville, "join our association and we will take you by the hand." "Upon what condition?" we inquired. "Oh, that you use and recommend our agents," was the reply. That is, if we would endorse their disease-making practice, we could have their protection and be a "scientific" doctor. If not we must expect their hostility and be called a quack.

The medical sect next lower in the scale of popularity is obliged to do a little more for the patient. Having less "scientific" prestige it is not allowed to make so much disease. As you go down the scale of popularity, you will find more and more simplicity, integrity and effectiveness in the treatment of disease. The reason is plain. Having no adventitious support in the shape of combined associatians, and not being able to humbug in the name of science, they are obliged to do their best.

If you want evidence of the retrogressive character of scientific medicine, examine the contents of the drug stores they mainly patronize. You will find their shelves filled with foreign drugs, no more necessary for the cure of our diseases than Japanese clogs are for the protection of our feet. Many of these articles, when they are unadulterated and of full strength, (which I am told is seldom the case,) are good medicines; but we do not need them. Our country abounds in roots and herbs of more value for the removal of acute or chronic disease than all the foreign drugs in the country. Ipecac, gamboge, senna, jalap, squills, rhubarb, colchicum, etc., etc., are all good medicines, but we do not need them. We have better, that cost only the digging and preparing. But scientific medicine so far as it could has ignored and contemned domestic remedies. A knowledge of their uses by the people would be very damaging to the profession. Even the foreign medicines named above are not extensively used by the "regular." They are too curative and unprofessional in their effects. "Scientific" medicine does not want medicines so much as drugs, chemicals, compounds of mercury, antimony, iron, bromine, arsenic, etc., etc. These are pets of the profession. With these the scientific man can doctor, and doctoring with these does in no wise injure the business, for the more they are used the more remains to do; and there is no fear of the people appropriating these drugs for domestic use; they experience enough of their disease-making effects in the doctors' prescriptions.

There is a tendency in sectarian medicine for healing arts to fall into disuse and forgetfulness. Many individual members, whose consciences protest against the use of disease-making minerals, are often a little reconciled to their use on the plea that they are required to use them by the association, and they cannot sustain themselves as independent practitioners. Many men have great facility in reconciling themselves to professional requirements that put money in their pockets.

As time passes doctors come to have an indistinct kind of faith in the absurd arts and practices that originated in sectarian interest. An observing mind will easily satisfy itself that there is nothing too extravagant to be believed if it only be enforced by a sufficient amount of authority. No doubt there are educated physicians who really believe calomel to be a medicine; and as faith increases in injurious and inutile drugs, real remedies are laid aside till a knowledge of their efficacy is lost or forgotten in the death of those who alone knew their virtue.

The history of medicine is not definite. It is impossible to form even a proximate idea of the sumber of medical sects which in different times and places have rose and reigned and fell. But it is certain that a great many have passed away without leaving a single enduring principle as a monument of their existence. Tongue cannot tell the number of disease-making drugs and appliances that one medical sect or another has introduced into the popular practice of the times. Commencing with mercurials there is a long list of mineral compounds that are utterly without value to the healer as curative agents. So far as their medicinal virtues are concerned, the whole catalogue is not to be compared in value with any one of a hundred plants that grow in abundance throughout the length and breadth of the land. Chemical compounds are the work of men's hands. Chemists make drugs to doctor with. It is only the Creator who makes medicines that heal. Chemistry has never originated a single medicine. It is not likely it ever will. If it were in the power of chemists to make real curative medicines, like those which abound in the fields, the profession would abandon the chemists; because their work would tend to destroy the profession. The most "elegant preparations" are inferior in acute disease to a simple decoction of crude leaves, root or bark.

Infinite wisdom has so planned the laws of organization that the healing arts are ever within the reach of the poor and humble, and almost as cheap as the rains and the dews of heaven. Riches and science have striven in vain to monopolize them, but the more they have clutched at them the more they have eluded their grasp. Innumerable modes of treating disease have been resorted to in different times and places. Vast numbers of these have passed into oblivion. Of those still in use none have stood the test of all ages but those involving the use of remedies found in the vegetable world, and those simple accessory appliances almost always within the reach of all individuals.

MEDICINE NO SCIENCE.

What is self-styled Scientific Medicine is a mode of practice in which the treatment of disease is pretended to be the legitimate result of scientific investigation. And such is the ignorance in regard to medicine that this absurd, misleading and false conceit obtains extensive endorsement by all classes of people. Only a superficial knowledge of medical art will convince any rational mind that no amount of scientific attainments will ever enable any practitioner to predetermine the appropriate treatment of disease. It is claimed that the medical use of Iron has a scientific basis. We showed in our article on Iron that no scientific analysis of the blood demonstrates a want of iron in that fluid. And that if in any case there was a deficiency of iron no personal appearance of an invalid will enable us to determine that fact Lastly, if we could be made to know the fact of its deficiency we do not know that iron, given as a medicine, assimilates and supplies that deficiency. We do know that when iron is given it removes no disease, but tends to fix, immovably, that which is in the system. Your Scientific Doctor has much to say about empirics and empiricism. An empiric is one that tries experiments; but every educated physician knows well enough that all there is or ever was in Medical Art of the least value is empiricism—experiment, and the results of experiment. Medicine has not been, is not, and never can be scientific; medical treatment is not even based on the deduction of science.

Medicine as a healing art consists in the *appropriate treatment of disease.* The attainment of any amount of Science does not determine that point. Chemistry, Anatomy, Physiology are Sciences, but a knowledge of them does not give us the slightest clue to the appropriate treatment of disease. Many men who have spent the vigor, vivacity and genius of youth in the acquisition of Sciences supposed to be at the bottom of medicine, who have devoted long years to the classification and description of disease, have sadly failed as curists or healers. Indeed they have more than failed; they have become skeptical in regard to the healing power of medicine just in the ratio of the learning they have exhausted. And this is much as if a pupil of Raphael should declare his want of faith in the beautiful and sublime art of painting, because, after acquiring much Science, he found himself unable to paint as well as his immortal master.

Medicine is an art depending upon knowledge obtained from two distinct sources. One of these sources is a knowledge of disease; the other, a knowledge of the effects and action of medical agents. A proper use of these knowledges can never be determined by Science.

A hot and dry skin—pain in the head, back and limbs, with the pulse at one hundred and twenty beats per minute, point to no remedy for the disease of which these are the symptoms and signs. Neither fever, inflammation, or any other form of disease carry with them any indications of cure. I know it is common for doctors in certain cases to say the indications are thus, or so. But this is only the judgment of the doctor, trained a little by previous experiment of his own or others. Science never has, never can help us here. Our knowledge in this direction is entirely empirical—purely the result of experiment. *When* medicine should be given; *what* medicine should be given, and *how* medicine should be given, are questions of Art, not of Science. If these points could be fixed by Science there would be but one mode of medical practice among civilized men, and to err in that would be in the highest degree criminal. There are different schools in Art, but there are no schools in Science.

Though Science does not enable us to predetermine appropriate treatment of disease, there is one way in which it may be used, properly enough, as a professional prefix. It is appropriate and truthful to say, scientific disease-makers. For it is easy to determine by Science what will make disease every time. And "scientific" doctors do agree in the uniformity with which they make disease. But when they try their hand at healing it is impossible to predict their treatment.

No amount of scientific study solves a single question of treatment. Even the most profound knowledge of disease implies no ability to treat it. There is no authority whatever for the treatment of disease but experiment. Scientific medicine, therefore, is a misnomer. He who claims to be a "scientific" doctor is either ignorant of the force of the language he makes use of, or is an impostor.

All medical sects wish to be thought scientific, just as hypocrites like to wear the livery of heaven. Science is used as a shield against criticism, and makes exclusiveness and intolerance plausible. But the true sons of Science are broadminded and tolerant. Agassiz asks not the aid of exclusive combinations, but quarries out the granite blocks of truth, unaided.

Self-styled scientific medicine, of all schools, is sectarian, and is aggressive, persistent and unrelenting in maintenance of the articles of its creed. Its dogmas, put in practice, make disease, and it always wants to monopolize the business of doctoring the disease that it makes. It refutes its own title by its perpetual strivings to compel opinion, and to crush out independent effort. No honest, conscientious physician, who earnestly devotes himself to the good of the public, will be allowed to get a living if it be in the power of organized "scientific" medicine to prevent him.

"You cannot succeed as an independent practitioner," said a "regular" to me one day. "What! not succeed, if I am really able to control disease?" I inquired. "No," said he "no matter how skilful you are there will be unseen influences that will nullify your best efforts. Everywhere detractors in the interest of the organized profession will belittle your best work—magnify and herald every little failure, and even censure that which most merits commendation. Your patrons will be persuaded to discharge you in the midst of your most successful labors, and ignorance and prejudice will be set in array against you. What! succeed because you are successful in treating disease? That will be the most substantial reason for crushing you out."

No one can doubt that "scientific" medicine does progress as civilization advances. But this progress consists in its increased facilities for imposition. It introduces new appliances which are only artful contrivances to increase professional business. It brings forward new agents which are not remedies, but new sources of disease; and it gives us new modes of treatment which are only new *Bugs Humming* the patient into slumber while the doctor empties his pocket.

"Scientific" medicine is a mode of treatment in violation of physiological law. If it seems to remove one disease it usually leaves behind it a greater. Real science ought to simplify appliances of treatment. But what is called Medical Science every way makes treatment more complicated.

No medical sect will admit any other mode of practice to be scientific. But each claims for itself what it denies to all others, which is sufficient proof that none are scientific.

HIGH SCIENTIFIC ATTAINMENTS NOT NECESSARY TO THE SUCCESSFUL PRACTICE OF HEALING ARTS.

True healing arts are simple. They depend upon an ability to judge of a few ill conditions of the human system, and a knowledge of the remedies for those conditions. Sectarian medicine, in order to keep the public ignorant, and to compel them to pay for disease making and protracting, has associated these simple arts with other arts, appliances and sciences, altogether irrelevant and of no practical utility to the healer.

Chemistry is a science of great value; but a knowledge of it is not essential to the practice of healing arts. If chemistry has furnished any aids to the practice of medicine, the physician can avail himself of them without the necessity of himself being a chemist. Healing arts, however, are not dependent on chemistry. They existed in their purity before chemistry was known. Essentially, chemistry has done nothing for the healing art. It has not produced a single agent that may truthfully be called a medicine. It has improved the form of some organic agents, and made them more convenient for use. It has not improved their action.

Great stress is laid upon a knowledge of anatomy as a basis for the practice of medicine. It is interesting to the medical student to know how to locate the principle organs of the body, especialy the visceral organs. It is a pleasing occupation to study the mechanical contrivance of the human system. It is a still more attractive and fascinating pursuit to acquaint oneself with the functions of different organs in the study of human physiology. But a knowledge of these sciences forms no part of the healing art, which is dependent on observation and experience or experiment.

Anatomy gives us no knowledge of disease. There is no disease in a dead body. Disease is a condition of the living organism. One might as well look among the ruins of a temple blown up by gunpowder for the purpose of ascertaining the elements and intrinsic qualities of that destructive compound, as to look into a dead body with the view of acquainting himself with the nature of disease. Disease can only be studied where it exists. If a dead body gives us no knowledge of disease, how can it possibly teach us the remedy for it ? A knowledge of anatomy is of service in surgical art, but surgery has necessarily no connection with medicine.

Although we are mainly indebted to the vegetable kingdom for healing agents, still a knowledge of the science of botany is in no wise connected with medicine. To be able to name, classify, and even describe the physiology of plants, is no guide to their medical virtues. And, contrawise, one may be familiar with the medical properties and uses of a multitude of plants, and not even be able to recognize them when growing in the field. The physician meets with most of his agents in the form which best serves the purposes of commerce. The study of their uses is entirely independent of their form, classification or physiology.

Sydenham and Laennec of the past, and Flint and others of our day, have devoted much time and study to the developement of an art for the detection of thoracic diseases. It is claimed that this art owes much to the discoveries of modern science. The value of the art, supposing it to be reliable, is greatly exaggerated. It is not of the least service in the treatment of the disorders which it is claimed that it enables us to describe. But, in truth, science has done little or nothing for this art. The signs upon which this theory is founded are furnished by the living, not by the dead. All of them, so far as they determine the presence of disease, are independent of anatomical knowledge, and may have served the physician a thousand years ago as well as they do now. Before it was known where the lungs in man are situated, or before it was known even that man had lungs. there is no reason why disease in those organs may not have been detected and successfully treated. We acquaint ourselves with disease by the study of its signs and symptoms, and not by the cutting up of dead bodies that have no disease. Inflammation and other affections of the lungs are attended by certain expressions that must have been regarded by the observing physician as characteristic of the presence of those disorders. These signs have been the same ever since man has had his present organization; and were as traceable at any time in the past as they are now by the light of all our science. We cannot dissect a man while living to examine his lungs; and if we could, they would tell us nothing of disease, but only the effect produced by it.

But this art of exploration, auscultation, percussion, etc., is very unreliable. Practically there is nothing in it but what might have been known, and probably was known to the ancients, and to studious, skillful physicians of all times. Whatever can be detected by the eye, the ear, the sense of touch, could always have been so detected.

Not many years since the writer was present, with other medical students, in a hospital of a large city, where several learned professors—former pupils of Flint —were exhibiting to the class what they called a case of phthisis, or tubercular consumption of the lungs. The patient coughed, and was greatly emaciated. A professor in a medical college, together with several other professors, percussed this patient with his fingers and also with a hammer and plate. He auscultated the patient with naked ear; that is, he studied the respiratory sounds by placing his ear to the chest of the patient. He also studied the respiration by the aid of a binaural stethoscope. He palpated the patient; that is, he tested the lungs by the use of his hands. He also endeavored to measure the inflation of the lungs by the eye. This he pronounced a clear case of tubercular consumption of the lungs. Four days later the patient died, and the writer was present to see the *post mortem* examination. And, reader, what do you suppose was found? Why, a pair of lungs in every respect perfect and sound. The man died with abscess of the liver.

The ready apology will no doubt be put forward that Science is no way to be blamed because these men made a mistake. No, Science *is* no way to be blamed, because there was no Science to blame. If one could determine the true condition of the lungs in disease by Science, no such mistakes could occur. This mode of judging of disease is only an art. and an art which is very unreliable, as we have shown, even in the hands of those who are thought its best representatives. The humbuggery and imposition practiced under the pretense of skill in this art are immense, and so gross that the veriest simpleton ought to be able to detect them. By this art any man may explore the chest and describe a supposed condition of the lungs. And who can refute his statements? We are not allowed to dissect our patients to see whether the doctors' diagnosis and prognosis were worthy of their confidence. So the whole thing resolves itself into a mere matter of opinion, and this opinion rules with the patient, not according to the science and skill of the doctor, but exactly in proportion to the authority that backs it. The professors who made the mistake, narrated above, may make such blunders every day, and their practice will not be injured by it, for such an expose of their ignorance and refutation of their diagnosis might not occur again in a lifetime. And if it did it would be likely to happen only among medical students, who, if they belong to any medical sect, would have a motive to conceal it.

A young doctor, fresh from the study of Sydenham and Laennec, and from the instructions of the polished Flint, was eulogizing the perfection of this art to an old physician whose early studies did not embrace this pet of the profession. The old doctor listened attentively until he had made an end, and then remarked in a squeaking voice, and in a style all his own. 'Yes, yes! God, God! but you can't see in there after all!!"

When the writer was attending medical lectures, a woman was brought before the class, laboring under a severe attack of inflammation of the lungs, for the purpose of enabling the students to study the disease. She was the subject of study for several days. Much was said over the case to enable the students to detect and describe the disease, but not ten minutes was devoted to the study or explanation of means that would relieve or cure. She was suffering terribly, and proper medical treatment would have relieved her in two hours. The question naturally arises, did the school devote itself to the promotion of healing arts, or was its main object to teach such arts as would enable its pupils to make most money out of disease?

The conservative principle of the human race requires that healing arts should not be dependent upon science, nor a high degree of civilization. The truth of this proposition is demonstrated in the fact, that the human race existed and flourished in population, in a savage or barbarous state, before science began to be developed. In regard to population and general purity of health, many barbarous and even savage tribes are quite on a par with civilized nations. Their number may be abridged by war and religious sacrifices, but not for the want of "scientific" doctors. No where among savages are the effects of malpractice so woefully manifest as in civilized countries where scientific practice is relied on by the deluded people. Does high civilization change therapeutic law? Not in the least. Inflammatory action in a savage is the same that it is in a philosopher. No special statistics are necessary to convince any well-informed rational mind that there is more disease; that it is of a more persistent and intractable character where "scientific" doctors abound than it is in those remote districts where the isolated inhabitants cannot avail themselves of medical science. Disease made by poisonous drugs is far more obstinate than spontaneous disorder.

No savage or barbarous races have declined and perished for want of scientific medical knowledge. They do much better without it than we do with it. They probably do not pay their doctor as much for making disease as for removing it. It is safe to say that we pay ten times as much for making disorder as we do for curing it.

POPULAR CREDULITY.

"I think I shall quit practicing medicine," said Dr. Williams, a "regular" friend of mine. "Will you tell me the reason?" I asked. "Well," said he, "the people are becoming so credulous that I am disgusted with professional experience. Formerly, if a man once got introduced into good families, he kept them as long as he wished. But now a man is sure of nobody. I find some that I thought my most reliable and sensible patrons running after healing mediums, clairvoyants, electropathies, Indian doctors, and every species of quackery. Talk about the intelligence of the people! I verily believe the world is going backward.

It was a *naive* and innocent conceit of the doctors, that they are retrograding, because they are leaving the calomel practice. They are really on the high road to progress, using their own faculties. perhaps for the first time, to find a better way. If the child falls, it is because he is trying to walk alone. He would never fall if he continued to lie in his cradle, nor would he ever learn to walk.

It is not those who are testing every new pretension that stand in the way of reform. They are indeed the real progressives. Do you tell me they get imposed upon and humbugged? possibly, but it is likely they have always been imposed upon. If they have not found it, they are, nevertheless, looking for a better way. It may cost them something to learn, but they are determined to be educated, cost what it will. There is another class that has not dared to question the immaculacy of the old doctors. These pride themselves upon being blindly wedded to impositions. Their stupidity is chronic. The old way is good enough for them. As their fathers did before them, they are contented to go to mill with the grist at one end of the bag, and a big stone to balance it at the other. If one tells them to divide the grain and so avoid the necessity of carrying the stone, they say, practically, that what was good enough for their fathers is good enough for them. This woman wants to be b'ed. "But, madam, bleeding is permanently injurious—you can be more effectually relieved another way." "I don't believe bleeding hurts any one," she said. "My mother was bled over fifty times. Bleeding didn't kill her; she died with dropsy." It would have been of no avail to tell this woman that dropsy is caused directly by bleeding. I referred to a man's blackened teeth and toothless gums as the work of calomel. He said his father had lost his teeth that way, and he did not know that he was any better than his father.

These are your sensible, firm people, who are not led away by every kind of doctrine. They are not to be humbugged by any new fangled ideas. Oh, no! they would like to see the tools they do it with.

This class hangs like a clog on the wheels of reform. The disposition of these people to be imposed upon is idiopathic, congenital, ineradicab'e. They may be bled till their blood is reduced to water. They may be calomelized till their teeth are destroyed—till their bones rot, and until they have not a sound tissue in their bodies, and they will still cling to this destructive practice. They never venture to criticise the treatment of a "regular" physician, but accept it as young robins do their food, with their mouths open and their eyes shut. They accept popular treatment even against the evidence of their senses. They have no opinions of their own, but are the very slaves of authority. The grossest malpractice, and the most skillful treatment, secure alike their approbation. All this comes from paying a blind devotion to what is mistaken for science.

You are an invalid. You want to get well. As a practical man, what have you to do with the acquirements of the physician? The best evidence of his ability is to be found in the character of his work. That is the best science for you that heals your malady. Here is the conceited Mr. A., whose pockets have been leeched for twenty years by systematic orthodox medical treatment. His own person bears the impress of its disease-making drugs, still he fancies he has a mission to warn the public against pretenders.

No man of common sense in these times has the least doubt that what is called Allopathic practice is a stupendous failure. In the most intelligent and investigating communities the percentage of those who are leaving it is greatest. Its maimed and crippled victims are seeking relief in every direction. You tell me that in many instances the people go to the merest mountebanks and medical hucksters. Well, that only demonstrates how desperate their case is—how entirely they have failed to get relief from their old doctors. The multiplication of new medical sects is proof positive that old ideas in that line are going out, because they have been weighed and found wanting. In the United States there is a vast population that stands confounded, amazed, disheartened by the outrages, cruelties, follies and absurdities of old school practice. They turn to the right and to the left for relief and faith. You who fancy yourself a public sentinel cry out through the iron collar of precedent you have worn for ages, "Beware! Beware! of these reformers. They have persuaded and turned away much people."

But this crowd does not heed you. These people have a suspicion that you are some way connected with one Demetrius, a silversmith, who has grown rich by selling little silver shrines and gods to the ignorant and credulous. They suspect that your solicitude about "imposition" arises from the fear that you will lose the monopoly of it.

Credulity! The people are not so credulous as regular doctors are skeptical, and it is but natural to expect that invalids will look otherwheres for relief from disease when what is called scientific medicine acknowledges its inability to cure.

REGULAR PROFESSIONAL SKEPTICISM.

Perhaps there is no fact better established than the skepticism of Allopathic physicians in regard to the curative power of their medicines. Many of the ablest writers of the calomel school have exhausted expression in the display of their skepticism of medical art—or as they vainly scall it, science. Dr. Dixon, of the New York Scalpel, says that "putting medicine into the mouth to cure disese is an absurdity. You can cure a man or a pig, but you can only do it when he is dead." A distinguished English physician calls medicine "the art of conjecturing." The celebrated Abernethy said, "disease is multiplied in proportion to the number of doctors." It is left for the reader to decide for himself whether the doctor doubted the power of medicine to cure, or the integrity and skill of his professional brethren. I think it was Dr. Good who said that "we know nothing of the effects of medicines except that they will kill." Dr. James Graham, alluding to the classical refinements, tinsel glitter and hifaulutin hypothesis of "scientific" medicine, says, "It hath in every age proved the bane and disgrace of the healing art." That is precisely what we trust we shall be able to show in these pages. Dr. Rush, of the University of Pennsylvania, says of "scientific" medicine, "Our want of success is owing to the following causes: First, our ignorance of the disease; second, our ignorance of a suitable remedy." It is impossible to read the more prominent works of writers on "scientific" medicine and not to be convinced that they all lack faith in their calling as a healing art. A volume might be filled with testimony going to prove this fact.

This skepticism in regard to the efficacy of medicine has been communicated to many otherwise intelligent people outside of the medical profession by the power of authority of prominent medical writers in connection with a knowledge of disease made by mineral drugs. The author of "The Poor Rich Man and the Rich Poor Man," puts in the mouth of a physician the following language, addressed to one of his patients: "Medicine has no more tendency to remove your disease than it would have to restore your leg if it had been sawn off and thrown away. Medicines, drugs, my child, are poisons. We are obliged to give them to arrest the progress of acute disease, but in chronic diseases they obstruct and clog the efforts of nature and confound her operations.

How it is that poisons must be given to arrest the progress of acute disease, while they aggravate chronic disease, the writer does not explain. The difference between acute and chronic disease is in most instances the same as that between a colt and a horse—one is older than the other. But this conscientious doctor says further: "I'll tell you a secret, my child; the older we doctors grow the less medicine we give." The constant use of mineral drugs is very apt to shake ones faith in the power of all medicines. While these drugs, themselves, have undoubtedly been introduced through the influence of exclusive organizations interested in making and protracting disease.

This want of faith is more characteristic of the denizens of cities and large towns than of the inhabitants of rural districts. There is more artificial life, and greater departure from nature, in cities than in the country. It would be as easy to make some people doubt the testimony of their senses as to shake their faith in the efficiency of medicine.

www.ingramcontent.com/pod-product-compliance
Lightning Source LLC
Chambersburg PA
CBHW022031190326
41519CB00010B/1664